Explore Space!

Space Robots

by Gregory L. Vogt

Consultant:
James Gerard
Aerospace Education Specialist
NASA Aerospace Education Services Program

Bridgestone Books
an imprint of Capstone Press
Mankato, Minnesota

Bridgestone Books are published by Capstone Press
818 North Willow Street, Mankato, Minnesota 56001
http://www.capstone-press.com

Printed in the United States of America.

Library of Congress Cataloging-in-Publication Data
Vogt, Gregory.
Space robots/by Gregory L. Vogt.
p. cm.—(Explore space!)
Includes bibliographical references and index.
Summary: Explains different types of space robots and their uses.
ISBN 0-7368-0199-5
1. Space robotics—Juvenile literature. [1. Robotics. 2. Robots.
3. Outer space—Exploration.] I. Title. II Series: Vogt, Gregory. Explore Space!
TL1097.V64 1999
629.4—dc21 98-045661
CIP
AC

Editorial Credits
Rebecca Glaser, editor; Steve Christensen, cover designer and illustrator; Kimberly Danger, photo researcher

Photo Credits
NASA, cover, 4, 6, 8, 10, 14, 16, 18, 20
Spar, 12

Table of Contents

NASA

Robots

Robots are machines that people program to do jobs. Robots can work in unsafe places. Robots can travel to faraway planets and work where people cannot go. Scientists use this robot to study volcanoes. They someday may use a robot like this on other planets.

planet
one of the large bodies that circle the sun

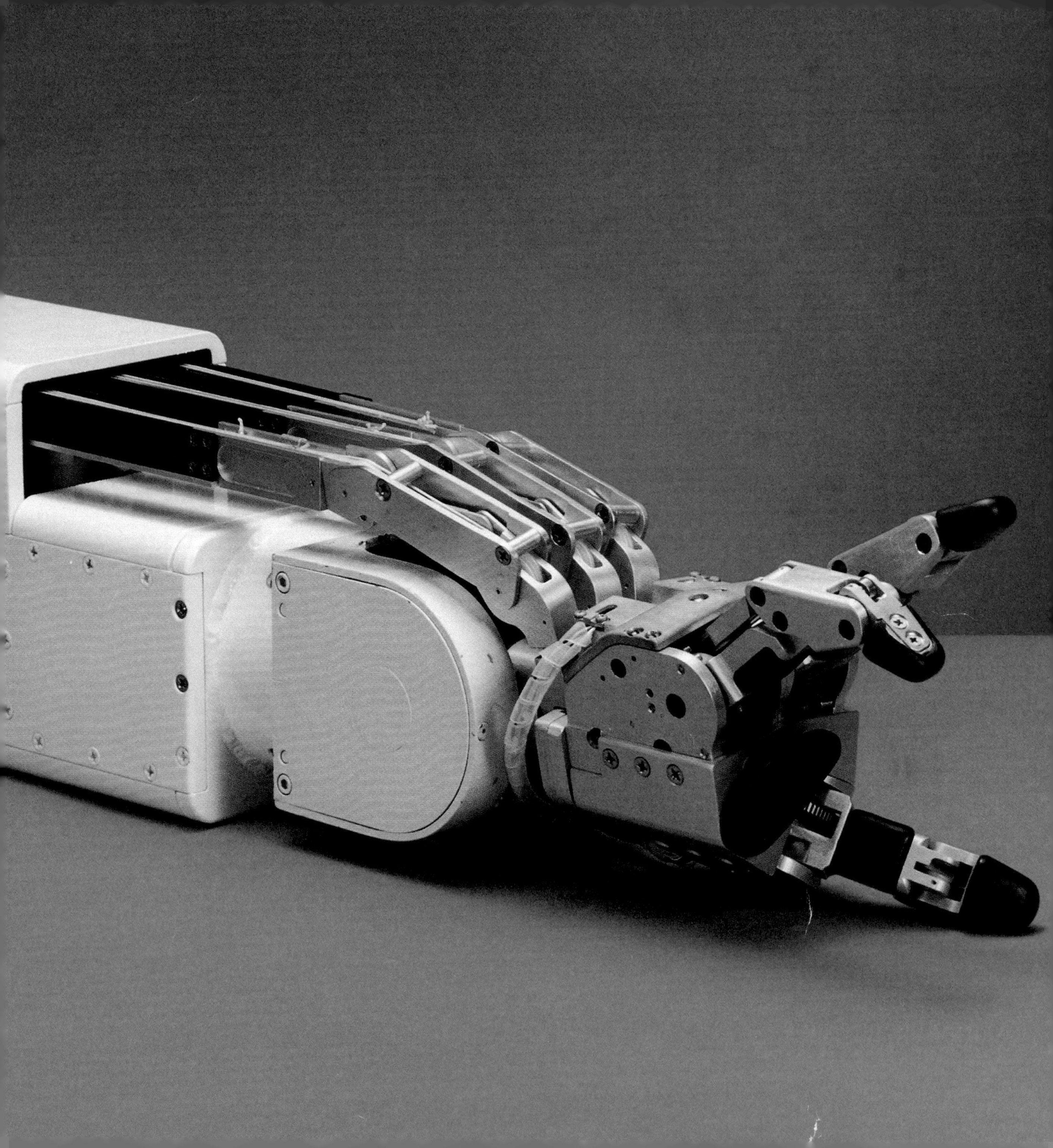

Robot Hands

Some robots have hands called grippers to pick up objects and tools. Grippers often are on the ends of robot arms. Most grippers have two to five fingers. Some robots have magnets or scoops for hands.

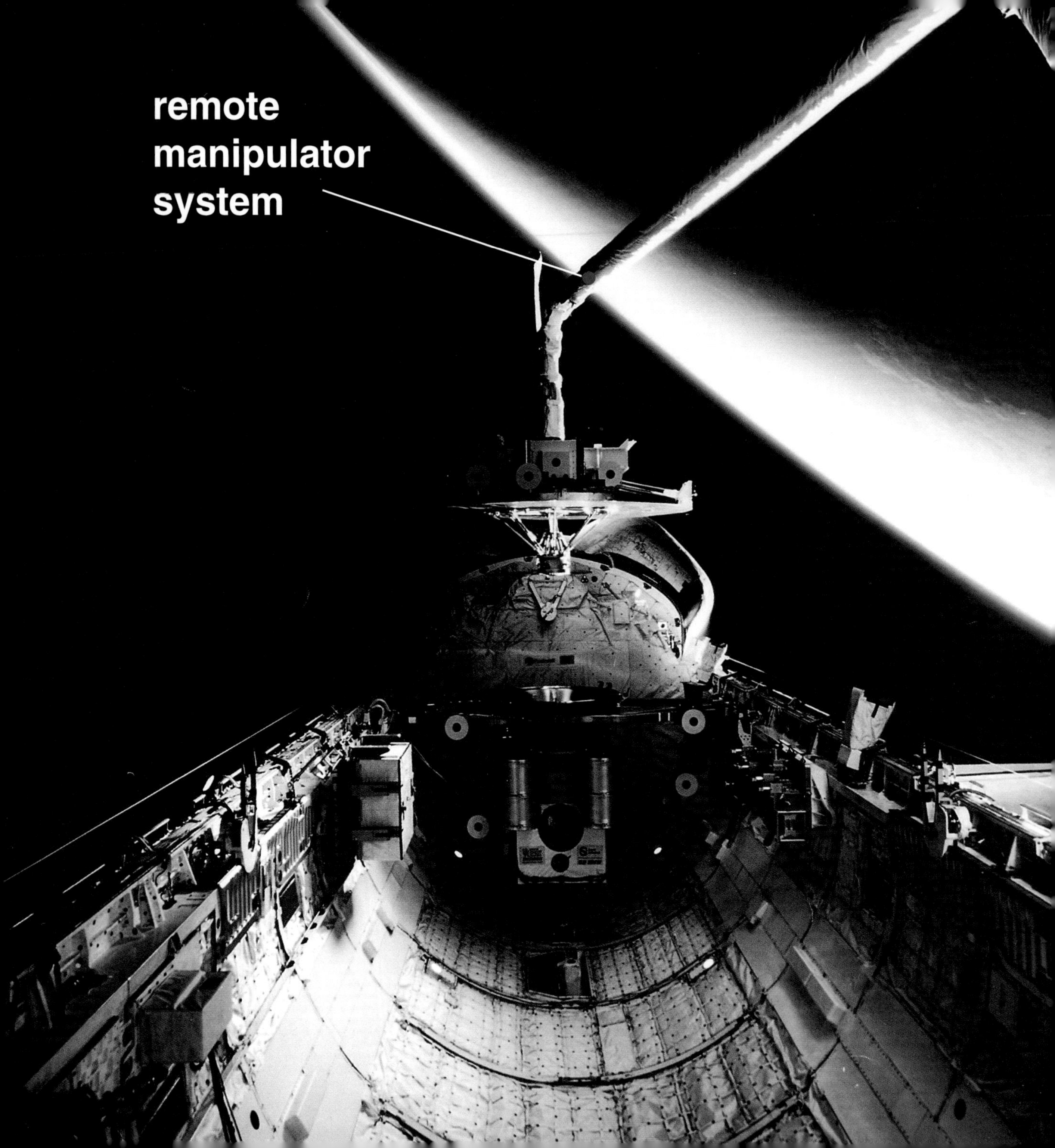
remote
manipulator
system

Remote Manipulator System

Astronauts on space shuttles use a robot called a remote manipulator system. This robot arm has a shoulder, elbow, and wrist. Astronauts use the robot arm to put satellites into space. Astronauts control the robot arm from inside the shuttle.

satellite

a machine that circles Earth; satellites take pictures and send signals to Earth and receive signals from Earth.

remote
manipulator
system

Working in Space

Astronauts also use the remote manipulator system when they fix satellites. Astronauts attach their feet to the robot arm. The robot arm moves astronauts to places that are hard to reach.

Canada

Space Station Robot

Scientists from many countries are planning the International Space Station. This space station will have a long robot arm with many joints. Astronauts will use the robot to build parts of the space station. They also will use it to move laboratories and make repairs to the station.

AERCam Sprint

Some robots carry cameras to places astronauts cannot see. The AERCam Sprint robot has TV cameras. Small rockets move this ball-shaped robot through space. Astronauts watch TV screens inside the space shuttle. They see what the AERCam sees.

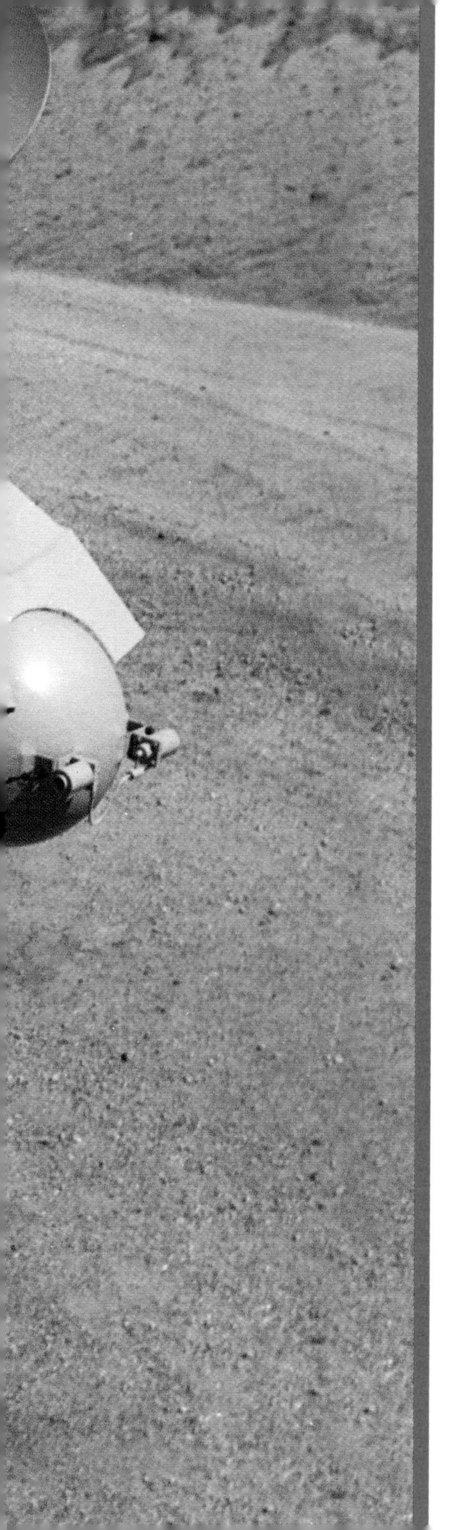

Viking I

Scientists use robots to explore other planets. Viking I is a space robot that scientists sent to Mars. Viking I took pictures of the planet's surface. The robot's arm scooped up dirt and studied it. Viking I also recorded the weather on Mars.

Sojourner

Sojourner is another robot that scientists sent to Mars. This small space robot was a rover. It moved across the surface of the planet to study rocks. People on Earth controlled Sojourner. They used a radio to tell the robot what to do.

Exploring Planets

Scientists use robots to explore planets far from Earth. Voyager II is a robot that traveled past Jupiter, Saturn, Uranus, and Neptune. Voyager II took pictures of these planets. It sent the pictures to Earth by radio waves.

radio waves
vibrations of energy that travel through the air

Hands On: Program a Robot

Robots are machines that do jobs people usually do. Robots need clear directions to do these jobs. You and a friend can try this game to see how robots work.

What You Need
One friend
One scarf or other blindfold
A book, ball, or other object

What You Do
1. Have your friend put on the blindfold. Your friend is the robot.
2. Give the robot directions to pick up an object. The robot cannot see. You must give clear directions like "Walk forward three steps."
3. Give the robot new directions if it does not find the object.
4. Switch places. You are now the robot. Your friend gives directions to you.

Words to Know

astronaut (ASS-truh-nawt)—someone trained to fly into space in a spacecraft

laboratory (LAB-ruh-tor-ee)—a place where astronauts do experiments to learn new things

magnet (MAG-nit)—a piece of metal that attracts iron or steel; some robot hands are magnetic.

space shuttle (SPAYSS SHUT-ul)—a spacecraft that carries astronauts into space and back to Earth

space station (SPAYSS STAY-shuhn)—a spacecraft that circles Earth in which astronauts can live for long periods of time

volcano (vol-KAY-noh)—a hole in the earth's surface; melted rock flows out of this hole when a volcano erupts.

Read More

Berger, Fredericka. *Robots: What They Are, What They Do.* New York: Greenwillow Books, 1992.

Gifford, Clive. *Robot.* Inside Guides. New York: D K Publishing, 1998.

Vogt, Gregory L. *Space Shuttles.* Explore Space! Mankato, Minn.: Bridgestone Books, 1999.

Internet Sites

CSA—Kool Zone

http://www.space.gc.ca/ENG/Kool_Zone/menu.html

NASA Space Telerobotics Program

http://ranier.oact.hq.nasa.gov/telerobotics_page/photos.html

Index